AF395624

ABOUT TIME

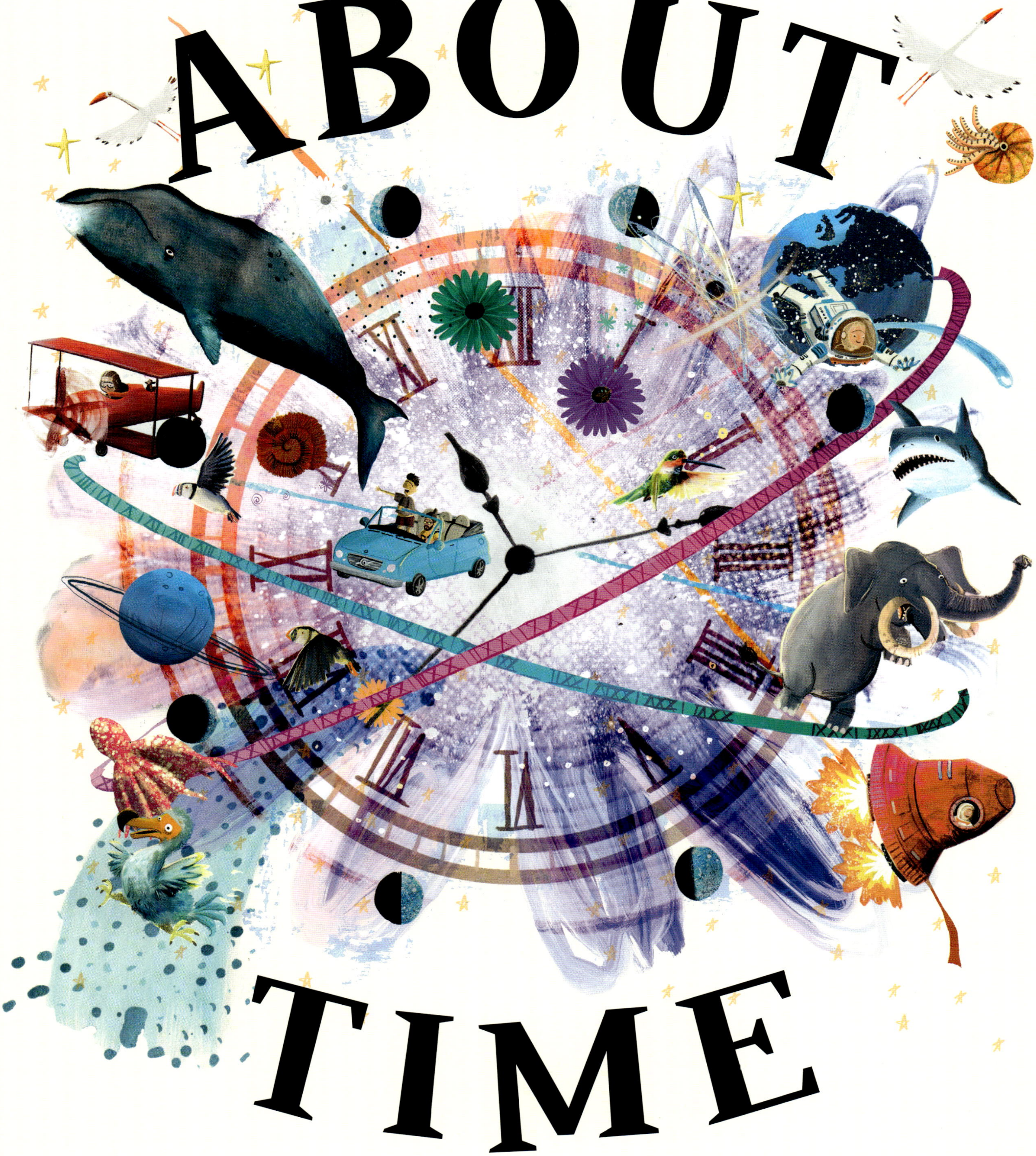

Published in 2024
First published in the UK by Barn Owl
An imprint of Igloo Books Ltd
Cottage Farm, NN6 0BJ, UK
Owned by Bonnier Books
Sveavägen 56, Stockholm, Sweden
www.autumnpublishing.co.uk
Copyright © 2024 Barn Owl

1024 001
2 4 6 8 10 9 7 5 3 1
ISBN 978-1-83795-940-2

Written by Tony Lee Moral
Illustrated by Aaron Cushley
Designed by Ashtyn Botterill
Edited by Katie Taylor

Printed and manufactured in China

At Bonnier Books UK, we are
committed to publishing sustainably.
Find out more here:
bonnierbooks.co.uk/sustainability

CONTENTS

HAVE YOU EVER WONDERED HOW TIME BEGAN?

It all started with the Big Bang! At the beginning there was nothing: no sun, planets, stars or galaxies – only pitch-black emptiness...

Within minutes of the Big Bang, the first atoms formed were the lightest gases: hydrogen and helium. 400 million years later, when the Universe began to cool, the lumps of hydrogen grew together and started to form balls of hot, glowing gas – the first stars.

13.8 billion years ago, our entire Universe was inside a tiny bubble, smaller than a pinprick. It was incredibly hot and dense inside, and it was ready to burst – a bit like a party popper packed with confetti. When the pressure got too much, there was a massive explosion known as the "Big Bang".

In a tiny fraction of a second after that superhot explosion, the Universe expanded by billions of kilometres in every direction. This created everything we see today, including planets, moons, stars and galaxies.

All the stuff around us grew from a big cosmic soup made of protons, neutrons, electrons and lots and lots of energy. These all joined to form atoms. Atoms are like tiny building blocks that make up everything in the Universe. They're so small that there are 60 million trillion in a single grain of sand! What's more, most of the atoms inside our bodies were once part of this cosmic soup, so we can say we are all made of stardust.

This is how our Universe was born, and it all happened faster than you could blink. It's how time itself was created.

There was an **explosion** of life. Our ancestors eventually appeared between 5–7 million years ago. These apelike creatures hunted animals and collected plants.

Over 1 billion years after the Big Bang, the first galaxies formed with trillions of stars and planets. Between 7 and 9 billion years after that, these new planets started to rotate around a star – our Sun – creating our Solar System.

Scientists believe *homo sapiens* (modern humans) evolved about 315,000 years ago.

Our planet, Earth, is about 4.5 billion years old – only a third as old as the Universe. Life first formed on Earth about 3.7 billion years ago, but this came in the form of tiny beings too small to see. Land animals only arrived about 500 million years ago!

Even though humans were late to the party, we can take a peek back in time at the entire history of the Universe simply by looking at the night sky and the fossil records on Earth.

HOW WAS OUR SOLAR SYSTEM FORMED?

When our Solar System was formed about 4.5 billion years ago, it started as part of a loose cloud of gas and dust known as the solar nebula.

The solar nebula was made of hydrogen, helium and lots of dust. Gravity started to pull the loose cloud together so it stuck in clumps. Then, the cloud of dust collapsed, possibly from the shockwaves from a nearby exploding star called a supernova. The result was a flat, spinning disk of dust and gas.

It was incredibly hot in the centre of the solar nebula, and the pressure there melded the gas, dust and other things (like metal from old stars) together to form our Sun. The rest of the spinning dust and matter clumped together through the force of gravity to form planets.

The Sun's strong gravity pulled the planets into its orbit, and they started to revolve around it like dance partners, spinning on their axes.

The number of days in a year on any planet depends on how close it is to our Sun. Planets closer to the Sun (like Venus) take a shorter time to orbit than those further away (like Neptune).

The cold planets made of gas and ice are further away from the Sun, so they take longer to orbit it. Saturn takes 29.4 Earth years, Uranus 84 years and Neptune 165 years.

The largest planet, Jupiter, takes almost 12 Earth years to orbit the Sun.

Mercury is the closest planet to the Sun and takes 88 Earth days to orbit it. Next is Venus, whose orbit takes 225 days. Being so close to the Sun means that these planets are too hot for anything to live there!

EARTH IS THE THIRD PLANET FROM THE SUN, SO IT'S NOT TOO HOT OR COLD.

Earth revolves around the Sun while rotating on its axis. This creates our day, night and seasons. Our calendar year is 365 days because it takes 365.25 days for Earth to revolve around the Sun. Over the course of four years, that adds up to a whole extra day, which is why we have a leap year. Adding an extra day in February keeps the calendar and seasons aligned.

The Moon is much smaller than the Sun, and it has less gravity. This is why astronauts have to take huge, jumping steps when they "moon walk" on it.

Our Sun is just one star in more than 100 billion that make up the Milky Way galaxy, which itself is just one of billions of galaxies that make up the universe.

There are more stars in the Universe than there are grains of sand on Earth!

The Sun's powerful gravity keeps the planets circling around it so they don't drift off into space. It's the same force that keeps us on Earth's surface so we don't float away.

WHICH PLANETS HAVE THE LONGEST AND SHORTEST DAYS?

A day is the length of time that it takes a planet to complete a full spin on its axis. This is what gives us our day and night. So, why is a day on Earth 24 hours long? It all comes down to the effects of the Sun and Moon...

When the Earth was formed, it spun much faster than it does today. A day on our planet was only 10 hours long!

Over millions of years, the Moon's gravitational pull has slowed down the Earth's rotation, causing the days to gradually lengthen at the rate of a millisecond per century. This has led to the 24-hour day we know today.

Of the eight planets in our solar system, Venus is the slowest rotating planet of all. It takes 5,832 hours to fully spin around — that's 243 Earth days! In fact, since Venus only takes 225 Earth days to orbit the Sun, one day on Venus is longer than an entire year on the planet.

A day on
the Moon lasts 29.5 Earth days,
so experts have been asked
to create an official time zone
for it so that astronauts can
tell the time while they're
up there.

The Moon slows our planet
right down because its gravity pulls
on the Earth's oceans. This creates tidal
bulges on the opposite sides of the planet
(making high and low tides) and acts like a brake
on the spinning Earth.

The Sun, being much bigger than the Moon, has an even
greater gravitational pull on Earth, and it speeds up
the rotation. It's like a tug of war between the Sun
and the Moon, with the Earth caught in the middle.
Our Moon spins on its axis every 29.5 days, which is
where the idea of our months come from.
Many cultures use 29- or 30-day
months to divide the year into
12 parts.

Do you ever wonder where the Sun goes once it has set?

Well, it doesn't actually go anywhere – it's the Earth
that moves. Different parts of the planet are either
facing away from the Sun or towards it. The Sun
only shines on the part of Earth which is facing it
and sunlight reaches different parts of the planet as
it rotates. As our location on Earth rotates into sunlight,
we see the Sun "rise", and when it rotates
away again, we see the Sun "set".

As for the other planets:
Mercury's day is 1,408 hours long
Mars's day is 25 hours long
Jupiter's day is 10 hours long
Saturn's day is 11 hours long
Uranus's day is 17 hours long
Neptune's day is 16 hours long.

WHY DO WE HAVE SEASONS?

We know that summer means long, fun days in the sun, barbecues and beach balls, while winter means shorter, cold days, snowmen and hot chocolate. But why isn't it summer all the time?

Rather than standing straight up, Earth is tilted at an angle of 23.5°. We think that something really big hit the planet a billion years ago and knocked it to its side. This made one part lean towards the Sun and the other part lean away. These are the **Hemispheres**.

As the Earth orbits, the Northern Hemisphere faces the Sun for part of the year, and at other times the Southern Hemisphere does.

When a Hemisphere points towards the Sun, the full strength of its rays shines directly on it, leading to hot summer temperatures. But when that hemisphere points away from the Sun, its rays fall on it at an angle, so temperatures are lower.

10

The Earth's tilt is why our seasons change as our planet orbits the Sun.

The day that the Earth's North Pole is tilted closest to the Sun is called the Summer Solstice, the first day of summer for the Northern Hemisphere. It falls in June and is when the North gets the most daylight hours, so it's known as the longest day. This is the Winter Solstice in the Southern Hemisphere, when the South Pole is farthest from the Sun.

The amount of daylight starts to reduce until the Winter Solstice occurs in December, when the North Pole is titled furthest away from the Sun. This is the shortest day of the year for the Northern Hemisphere. Meanwhile, in the Southern Hemisphere, the South Pole is closest to the Sun, so it's the Summer Solstice there.

During spring and autumn, temperatures are mild because the hemisphere neither faces nor slants away from the Sun's rays. If the Earth didn't rotate, one side would always be hot and the other would always be cold.

When early Mesolithic hunter-gatherers built Stonehenge around 5,000 years ago, they aligned the huge stones with the Sun so they could keep track of the midsummer sunrise and midwinter sunset.
It was like a huge, ancient calendar!

HOW OLD IS EARTH AND HOW WAS IT FORMED?

Our planet is the only one we know of that contains life.

This makes Earth very special, as it had just the right conditions for the spark of life to start. And it all started with an exploding star...

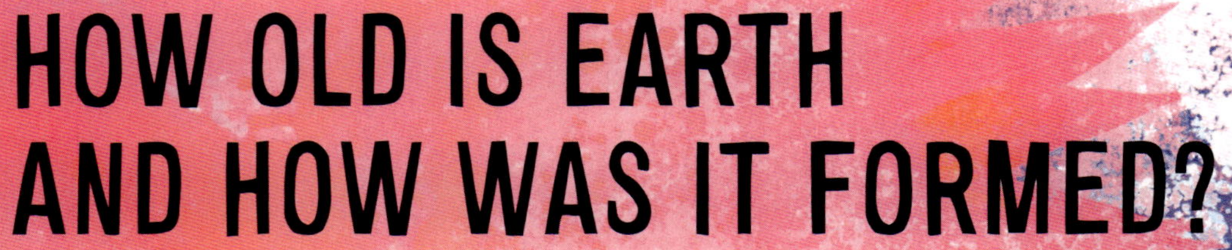

After the Sun was formed, there was lots of dust and twirling rock left over, spinning in the solar nebula. Over time, it clumped together and built up our Earth. Giant meteors of ice and rock hit the planet, making it grow bigger and bigger until it became a giant ball of melting hot rock.

On the young Earth, volcanoes erupted all over the surface, spewing hot gas and lava. As the planet cooled, water formed from the meteors of ice and gas that had smashed into it. Water also rose through the volcanoes to form steam clouds, which then rained down to form our oceans. Now, over two-thirds of the Earth's surface is water, which is why we call it the blue planet.

A really big, hot rock smashed into Earth, and a huge chunk tore off and floated into space. Gravity kept it within Earth's orbit and it became our Moon. We always see the same side of the Moon because it takes the same amount of time to rotate on its axis as it does to orbit the Earth.

Volcanoes also erupted underwater. The molten lava formed volcanic islands, which joined together to form a giant continent called Pangea. Over millions of years, this supercontinent broke up and drifted apart to form the smaller continents we see today.

What do you think Earth will look like 100,000 million years from now?

OVER MILLIONS OF YEARS, THE TOP LAYER OF THE EARTH COOLED AND HARDENED, AND THE SURFACE BECAME LIKE A CRUST OF BREAD.

Plants started to grow on land about 500 million years ago.

The first animals crawled onto land about 574 million years ago.

Fish appeared in the sea about 530 million years ago.

The first life eventually formed on Earth about 3.7 billion years ago in the shape of microbes.

WHAT ARE THE FOUR ERAS OF EARTH?

Have you ever looked at the rock beneath your feet and wondered how old it is?

We can measure how old the Earth is through a geological time scale. Older rocks are lower down in the Earth's core while newer ones are at the top. Each layer of rock can contain fossils or remains of living things, which are like a geological clock that can be divided into four main eras: Precambrian, Palaeozoic, Mesozoic and Cenozoic.

The Precambrian era is the oldest era of geological time, spanning from 4.6 billion–541 million years ago. It's where early life started around 3.7 billion years ago. The first living things to appear were tiny, single-celled bacteria. In the sea, creatures with squidgy, soft bodies evolved, like worms and jellyfish.

Next came the Palaeozoic era, which was 541–252 million years ago. Palaeozoic means "ancient life", and during this time there was a huge explosion of it, marking a major milestone in evolution. Many different plants, fish and reptiles evolved (mostly in the sea) and there were lots of strange-looking creatures, like ammonites, which had swirly shells a bit like snails.

Precambrian 4.6 billion– 541 million years ago

Palaeozoic 541–252 million years ago

However, the era ended with the biggest mass extinction ever seen. Changing climates wiped out 95 per cent of marine life and 70 per cent of life on land. This may seem like a lot, but it made room for new kinds of animals to evolve.

On land, the Earth was hot and full of deadly gases like methane, but over time, oxygen started to build up in the atmosphere with the first algae. The Precambrian is the longest geological era, taking up 88 per cent of time on Earth.

The Mesozoic era was 252–66 million years ago. At this time, the supercontinent Pangea was starting to break apart and spread out. Much of Earth's climate was lush and tropical, and the age of reptiles was born.

The last era is known as the Cenozoic, which means "recent life" and stretches from 66 million years ago to the present. It is the final time period on the geological time scale. The few creatures that survived the deadly meteor, such as small furry mammals, became Earth's main inhabitants. Flowering plants, insects, and all life as we know it today evolved into its present form.

The most famous Mesozoic creatures were the dinosaurs, which ruled the Earth for over 150 million years. The era ended with their extinction, when a giant meteor the size of Mount Everest smashed into the planet and killed off 75 per cent of life on Earth.

Humans are the last page of the story so far in the Earth's time scale, popping up a few million years ago. If we imagine the history of the Universe as one year in the cosmic calendar, with the Big Bang on January 1st, humans appeared at 30 seconds to midnight on December 31st.

THE PERMIAN ERA

Over 250 million years ago, long before the dinosaurs or mammals appeared, Earth was dominated by some bizarre-looking creatures: the Permian monsters...

The Permian era is the final period of the Palaeozoic. Pangea stretched across the planet and hosted animals with sharp teeth, like Gorgonopsians, which evolved 12-cm-long fangs and were the world's first sabre-toothed carnivores.

In the sea there were strange looking sharks and bony fish with thick scales and fan-shaped fins.

The first conifer trees appeared, and the prehistoric forests were crawling with bugs – but not like any you've seen before. The Permian was a time of giant insects, massive millipedes, and dragonflies the size of crows. Insects grew so big because there was more oxygen on Earth than there is today.

One very famous Permian beast was Dimetrodon, a 4-metre-long carnivore with a huge sail fin on its back and a long tail. The sail fin was thought to act like a solar panel to catch the warm sunlight, allowing them to warm up quicker in the early mornings.

There were reptiles called Cynodonts which were covered with hair, suggesting they were warm-blooded. They later gave rise to the mammals.

The Permian era ended with a mass extinction, where over 90 per cent of life on Earth was wiped out. The planet became very hot which was caused by erupting volcanoes and led to changes in the ocean. All of this made way for the next Earth rulers: the dinosaurs...

THE MESOZOIC ERA

The Mesozoic era is divided into three periods: the Triassic, Jurassic and Cretaceous. It's best known for being the age of the dinosaurs!

There were many different dinosaurs around. Over 700 types have been found so far!

During the Triassic, between 252 million and 201 million years ago, dinosaurs lived on the single, giant continent called Pangea, and the climate was warm and dry. Most dinosaurs were small predators, such as Coelophysis. This dinosaur could grow up to 3 m long. It had hollow bones and was fast and agile, using its speed to catch small reptiles and insects.

The Jurassic Period was between 201 million to 145 million years ago, and was a golden time for dinosaurs. Plant life flourished and Pangea broke up into separate continents. Dinosaur fossils have been found on every modern continent including Antarctica, which tells us that the dinosaurs spread across the Earth.

The biggest dinosaurs were the Titanosaurs, which could reach over 26 m long and weigh as much as 12 full-grown African elephants. They belonged to a group called the sauropods, which include famous dinosaurs such as Diplodocus and Brontosaurus. Despite their huge size, they were gentle plant-eaters, grazing on the tops of trees.

During the Cretaceous Period (145–66 million years ago), Earth hosted some of the most famous dinosaurs, like T. rex, Triceratops, Iguanodon and Spinosaurus.

T. rex dominated the Cretaceous as a top predator. It could reach 13 m long and weighed 9 tons – as much as two hippos. Its heavy skull contained 60 cone-shaped teeth, each as big as bananas, making T. rex one of the biggest and scariest hunters around.

The reign of the dinosaurs ended abruptly when an enormous meteor crashed into what is now the Yucatán Peninsula in Mexico. Its impact created a gigantic crater, sending dust clouds into the sky and blocking out the sunlight. Without food, the plant-eating dinosaurs died, followed by the starving meat-eaters.

The Mesozoic era starts with the evolution of the dinosaurs and ends with their extinction. But some small animals survived, and they gave rise to the mammals...

MONKEYING AROUND

Humans are related to all other animals, but we are especially close to our primate cousins: apes and monkeys.

Although we share a lot of our genes with chimpanzees, we didn't directly evolve from apes, but from an ape-like ancestor. If we go back in time to 45 million years ago and look along our family tree, we'll find an early type of monkey called *Eosimias*. It was so small it would fit in the palm of your hand! It was from this common ancestor that all primates evolved, including us.

Our evolution is like the branches of a tree, with our human ancestors on one branch and the apes on another.

Historians believe that man's early ancestors broke away from the apes 7 million years ago and started on their own branch of evolution.

Our ancestors evolved through a series of more than 20 different early humans (hominoids)! Researchers have found they used stone tools to hunt food, with evidence dating back to between 2.6 and 3.3 million years ago.

One of the earliest hominoid fossils found was a female *Australopithecine*: an ape-like human whom scientists called Lucy. Lucy lived in modern-day Ethiopia about 3 million years ago. She stood just over 1 m tall and could walk on two legs.

Another of the early humans was *Homo erectus*, which means "upright man". *Homo erectus* had a large brain and small teeth, and hunted animals. They began to make tools out of sticks and stones, and about 1 million years ago *Homo erectus* discovered how to make fire, which they used to cook food and keep warm.

As man evolved, our brains became bigger. Then, 300,000 years ago, *Homo sapiens* (which means "wise man") evolved. This is the species we are today!

Time has made us the smartest animal on Earth, and we have shaped the world like no other living thing before us. Humans are still evolving, developing the latest smartphones and firing rockets into space. What could we do next?

THE STONE AGE

The Stone Age began about 2.6 million years ago. It was the earliest form of human culture, when early man started developing tools. The Stone Age is divided into three periods: the Palaeolithic, the Mesolithic and the Neolithic.

The Palaeolithic is the oldest and longest period. It started with the first humans and lasted until 10,000 years ago.

During this time, early humans chipped flint stones and rubbed two sticks together to create a spark and light fires. Our early ancestors ate fruits, berries and honey, and started to cook their meat.

Weapons were made out of stones, which is why we call this era the Stone Age. Handyman skills were developed when early man made chisels and scrapers with handles attached.

The Neanderthals were a type of Palaeolithic human who lived in Europe and Asia during the Ice Age.

They looked a bit like pro wrestlers, as they had small, muscular bodies with big noses to heat up the air they breathed during the harsh winters, and wore clothes made of animal fur. They used arrows and spearheads for hunting and fishing.

The last Neanderthals died out about 40,000 years ago, but *Homo sapiens* continued to evolve.

Did you know that
the Neanderthals were musicians?
A flute made of bear bone was found
that dates back to their time. It's the
oldest musical instrument we know of!

Next came the Mesolithic period, which means
Middle Stone Age. Our ancestors started to build
basic shelters from wood and covered them
with animal skins.

This was followed by the Neolithic, which means
New Stone Age. Man became more skilled,
using mud to make pottery and bricks for houses.
We started to live together in villages.

People learnt to grow crops and began farming.
They developed a range of tools and started breeding cattle.

At the end of the Stone Age, people began making copper tools, which led to the Metal Age. This era is made up of the Copper, Bronze and Iron Ages, and was the first time that humans began to use metal instead of stone tools. Copper helped the Neolithic people develop new crafting skills, which led to...

THE BRONZE AGE

The Bronze Age took place between 2300–700 BCE. During this time, humans learnt to make very hot fire, and used it to remove bronze and iron from rock in a process called metallurgy. Bronze soon replaced stone as the preferred choice for making tools and weapons, and led to new inventions like bronze axes and swords.

The Bronze Age was the last prehistoric period for some civilisations because the invention of writing meant that their history could be officially recorded, giving us an idea of what happened back then.

The wheel was also invented during the Bronze Age, at first used as a potter's wheel to help shape clay. The earliest wheeled vehicles appeared around 5,300 years ago. The Sumerians (who lived in what is now Iraq) used them to make it easier to trade and carry goods. The Sumerians were a busy bunch – their maths skills are the reason we count 60 seconds per minute and 60 seconds in an hour.

Our Bronze Age ancestors lived in roundhouses made of wood and stone with a thatched roof. They were quite a fashionable crowd, using wool to make kilts and cloaks. Men wore long tunics and women wore woollen skirts.

THE IRON AGE

The Iron Age began in the Middle East in about 1200 BCE, and lasted around 1,000 years. This era saw the first use of iron to make tools and weapons.

The Celts lived in places such as Britain, Ireland, France, Germany and Spain. They were fierce warriors, with weapons and armour (like helmets and shields) made of iron, which was stronger than bronze.

People lived in hill forts, surrounded by walls and ditches. Inside the forts, round houses were plastered with clay and straw.

Iron Age people were also farmers. They invented the metal plough so they could grow and collect more crops all year round. They also kept cattle, sheep and pigs.

The Romans marked the end of the Metal Age in the 1st century BCE when they began their conquest across Europe.

THE ANCIENT AGE

The Ancient Age is the time between the invention of writing and the start of the Middle Ages.

It started around 5,000 years ago and lasted until about 476 CE. It is generally marked by the rise and fall of great empires such as the ancient Egyptians, ancient Greeks, Mesopotamians, Persians, and ancient Romans. They built the first cities and gave rise to great civilisations, with the invention of writing, pottery, and use of metals.

It was an influential time as trade and cultures were developed and exchanged. Weapons were developed for the many wars that happened during this time.

The Ancient Age ended with the fall of the Western Roman Empire.

What we think of as ancient Egyptian civilisation began in the late Stone Age 5,000 years ago and lasted for 3,000 years. Architectural wonders were built out of stone, like the Great Pyramid of Giza, which is over 228 m wide and 146.5 m high. The pyramids were built as giant monuments for Egypt's pharaohs, who were buried inside.

The Roman Empire stretched across Europe. Its capital, Rome, was perhaps the greatest city in the ancient world. People gathered in colosseums to watch gladiators or wild animals fight to the death.

The first people to use writing were the Sumerians and ancient Egyptians. It started in modern-day Iraq in the form of an early writing called cuneiform, where letters or characters were carved into soft clay tablets using tough reeds. Soon, different cultures across the world were using different characters and symbols to create the written word.

The pharaohs had their treasure buried with them as they believed it would help them in the afterlife. When the tomb of the young King Tutankhamun was discovered in 1922, it contained lots of treasure, including three golden coffins and a 10 kg golden death mask.

EARLY CLOCKS

The first proper evidence of timekeeping dates back to around 5,000 years ago, when humans were farming. These early farmers used the Sun, Moon and water to tell the time, and they started to use their understanding of nature to build some basic clocks.

Candle clocks use fire, and have been used around the world from China to England. They are thin candles with equally spaced markings on the side that show the passage of time, and don't rely on sunlight. In the 9th century, King Alfred the Great used six 30-cm candles that each burned for four hours to tell the time, making a 24-hour day.

The ancient Egyptians made the first shadow clocks about 3,500 years ago to measure the passing of time during the day. They were called obelisks, and were tall, stone monuments with a small pyramid at the top. Later on, miniature obelisk clocks called sundials were created.

The stick that casts the shadow on a sundial is called a gnomon. When the sun passes through the sky, the gnomon's shadow moves across the ground. The Egyptians divided the day into 12 hours and used the position of the sun to measure time by tracking its shadow. The problem is, you can't use sundials when it's cloudy or at night.

The hourglass is another primitive type of clock. Here, fine sand pours through a small hole at a steady rate to show the passing of time. Hourglasses were especially used by 14th century sailors to help them navigate. When the Portuguese explorer Ferdinand Magellan travelled around the globe in the 1500s, he had 18 of them on his ship!

The water clock is another of the oldest timekeeping devices used in Egypt, Persia, India and China. Water fell from one bowl into another, dripping at a steady rate. Users would count how many drops fell in a minute, while markings on the side of the bowl showed what the time was. Water clocks do have their drawbacks though, because water freezes in the winter and evaporates in the summer, making seasonal time-telling unreliable.

Did you know the world's first alarm clock was invented in ancient Greece? The philosopher Plato developed a kind of water clock where liquid built up in a jar and eventually forced air through a whistle to wake him up.

MODERN CLOCKS

Modern clocks were a big change from hourglasses, water clocks and sundials! These ones relied on a swinging pendulum or a vibrating quartz crystal, and were much more reliable timekeepers.

The first mechanical clocks were invented in China in 725 CE, with more following in 13th century Europe. They were huge devices, driven and powered by strings and a hanging weight. These early clocks had only a single hand to mark the hour at first, with minute hands appearing around the 1600s.

By the 17th century, pendulum clocks had been developed where free-swinging weights hung from a pivot. The use of pendulums to accurately measure time was discovered by the Italian astronomer Galileo Galilei. He noticed that pendulum beats always took the same length of time, no matter what distance the pendulum swung. It was the most precise timekeeper in the history of clocks up to that point.

Many countries and militaries use the 24-hour clock, or "military time". Instead of saying "1.00 p.m." after 12 noon, military time keeps counting, making it 13.00, 14.00 (2 p.m.), all the way up to 23.00 (11 p.m.) and 00.00 (midnight). This is to avoid confusion between morning and night.

Pocket watches were invented by a German locksmith in the 16th century. They were portable clocks with a balance spring inside instead of a pendulum. Charles II, a 17th century king of England, loved them so much he had a pocket in his fancy waistcoats just to keep his pocket watch in.

All of these inventions have meant that our timekeeping has improved. Today we have very accurate clocks that measure 24 hours, 60 minutes and 60 seconds with digital displays.

The first spring-driven pocket watches were developed in the late 14th century. They were used during World War One but were considered impractical in combat, so soldiers and airmen started wearing wristwatches. This made it much easier to tell the time during the heat of battle!

By the 1930s, pendulum clocks had been replaced by electric clocks. This change was accelerated when the quartz crystal was introduced in 1929. When the crystal is charged with electricity, it vibrates. This controls a motor and drives gears to accurately move the clock hands.

WHY DO WE HAVE TIME ZONES?

Have you ever wondered why different places on Earth have different times?

It's all because Earth is divided into time zones, which were created because of our planet's rotation! The time differences are because the Earth is a round, spinning ball, and some cities are in opposite time zones. For example, when it's 11 am in London, it will be 8 pm in Sydney the same day. In Beijing, someone may be having their dinner at 7 pm, whereas New Yorkers will still be just getting up at 7 am.

20 other countries agreed to divide the world into 24 time zones, one for each hour of the day. The Earth was divided into imaginary "longitude" lines running vertically north to south, each 15° apart. The Royal Observatory in Greenwich, London, was chosen to be the starting point of 0° because it was trusted by travellers for its reliable timekeeping. This 0° line is called the Prime Meridian and it splits Earth into east and west. The time here is Greenwich Mean Time (GMT), and the time around the world is compared to hours ahead of (or behind) it. As you move west from London, every 15° section is an hour earlier than GMT, and as you move east, it's an hour later.

It was decided that the USA needed to have one timekeeping system so things didn't get confusing. They liked how the UK railway used one standard time and decided to introduce it in America in 1883. Since the USA was so large, it was divided into four time zones: Eastern, Central, Mountain and Pacific times.

Although we haven't invented time travel yet, we can go back in time by flying westwards across the Atlantic Ocean. You could take off at 11 am in London, fly for eight hours, and it would only be 2 pm when you land in New York.

OTHER WAYS OF TELLING TIME

Science has found some clever ways to tell the time which don't rely on clocks.

Trees are like living time machines! The rings in their trunks not only tell us how old the tree is, but also what the weather was like at the time – if there were any cold spells, for example. By examining them, we can build up an amazing picture of the Earth during their lifetime. This science is called "dendrochronology".

It's not just trees that have their own built-in clock; fossils do too. We can measure the amount of carbon-14 (a chemical element) that remains in a fossil through a process called radiocarbon dating.

Carbon-14 is naturally found in the air, and plants and animals absorb it as they breathe. Around every 5,730 years, half of the carbon-14 atoms will decay into nitrogen. This is known as the "half-life," and allows us to work out how many thousands of years have passed since the animal or plant has died by measuring the amount of carbon-14 in its fossil.

In a similar way, we can date rocks through a technique called uranium–thorium–lead dating, which works by measuring the amount of lead compared to the amount of uranium a rock has. It doesn't stop there, though — even soil can be dated by luminescence dating. This measures the amount of light emitted from energy stored in soils, as well as quartz, diamond and calcite.

We can also tell the time by looking under our feet. Many modern roads, for example, follow routes which are hundreds or even thousands of years old. Old cobbles and even fossils and treasure underneath the tarmac can tell us how old a route is.

Ice cores are long pieces of ice taken from a glacier. The deeper the ice, the older it is. We can date the ice by looking at its layers to find out what the climate was like. The colour of the layers tells us if gas, particles and ash were spewed up from erupting volcanoes.

CAN WE TIME-TRAVEL?

Would you want to go back in time?
Or fast-forward to the future?
Is it even possible to make a time machine?

It sounds like the stuff of science fiction, but the fact is we're all travelling in time at the rate of one second per second. But can we actually time-travel?

The faster you travel, the slower you experience time. Clocks on aeroplanes and satellites travel at a different speed than those on Earth. An experiment showed that a clock on a plane was travelling slightly slower in time than one on the ground.

It's possible to experience time passing at a different rate, so if we travelled really fast (up to even half the speed of light), we might – in theory – time-travel into the future..

Over a hundred years ago, scientist Albert Einstein came up with his theory of relativity. He said that time and space are linked, and that nothing can travel faster than the speed of light (186,000 miles per second). Einstein saw time as a fourth dimension.

In 1935, Albert Einstein also proposed the existence of bridges or shortcuts in space. He gave them the name "wormholes" and theorised that they could be portals through space and time. Perhaps time travellers in the future could use these wormholes to travel through the vortex!

Did you know that astronauts go forward in time when they go to space? Those on the International Space Station also age slightly slower due to the spacecraft's high speed.

Time travel has been theorised a lot. Most experts believe it is not possible to go back in time, as what has happened in the past cannot be changed. However, since it takes about 13 milliseconds for our brains to process light, we are actually viewing something that has already happened, so we are in fact always looking at events in the past.

It takes a long time for the light from faraway galaxies to reach us, so when we look through a telescope, like the James Webb Space Telescope, we're actually seeing what those stars looked like a very long time ago – literally looking back in time!

TIMELINES

Timelines help us understand when events happened in relation to each other.

Did you know that Tyrannosaurus rex lived closer to us humans than it did to Stegosaurus? T. rex lived in the Cretaceous Period 65 million years ago, whereas Stegosaurus lived 150 million years ago during the Jurassic Period.

Humans have been around for 300,000 years but that's a very short time compared to the Earth dinosaurs, who ruled the Earth for 160 million years!

The oldest living tree is a bristlecone pine in eastern California, which is over 4,850 years old according to its ring data. The tree, known as Methuselah, is Earth's oldest living thing – it was already 1,000 years old when the last mammoth died!

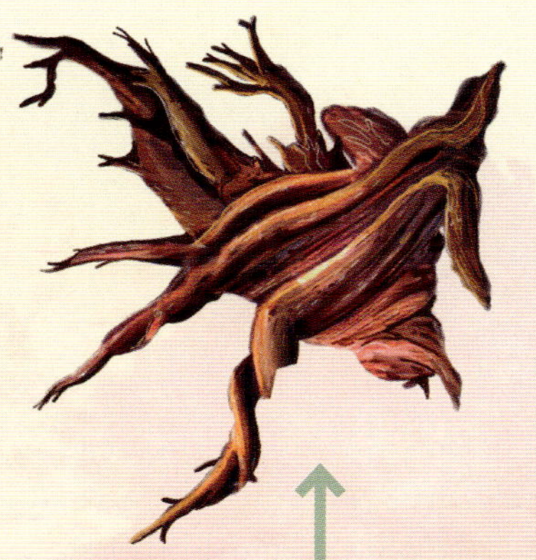

The pyramids were built without wheels, so the 100,000 people who built it had to haul the big slabs of stone themselves on sleds. They constructed it using stone and copper tools.

The Egyptian queen Cleopatra lived closer to the invention of the first smartphone than to the building of the first pyramid at Giza in around 2,560 BCE. That's because she was born in 69 BCE and lived until 30 BCE.

The distant past...

BCE

243,000,000 BCE	150,000,000 BCE	4000 BCE	2800 BCE	2560 BCE	1650 BCE	31 BCE

TIME JUMP!

The Sphinx in Egypt was built around the same time as Stonehenge in England, as both were built around 5,000–4,500 years ago.

Woolly mammoths were alive in 2560 BCE when the pyramids were built. Most died out 10,000 years ago, but a few survived in Russia until 1650 BCE, when the pyramids were already 1,000 years old.

CE

1250 CE 1330 CE 1800 CE 1850 CE 1860 CE 2006 CE Present Day

The first dinosaur fossil was discovered in 1824. George Washington died in 1799, so he never knew that dinosaurs existed!

The London Underground was opened in 1863 while the Civil War was still being fought in America.

Pluto was discovered in 1930, but was declassified as a planet in 2006, which means it didn't even get to complete one orbit (which takes 248 Earth years) between its discovery in 1930 and its downgrading.

The UK's Oxford University dates back to 1096 CE, which means it was already around when Mexico's Aztec Empire arose n 1325 CE.

Some whales alive today are older than Moby Dick, the famous whale in Herman Melville's novel. Bowhead whales can live to be over 200 years old and Moby Dick was published in 1851.

Harriet the Tortoise, who died in 2006 at the age of 176, was alive when Charles Darwin visited the Galapagos Islands in 1835.

ANIMAL AGES

Some animals have very short lives, while others live long enough that they were born when our great-grandparents were alive. There are even a few that are very close to living forever and can be considered immortal!

Larger animals generally live longer than smaller animals. Their hearts beat slower and they have slower metabolic rates (the amount of energy an animal uses).

Scientists have found special genes inside the cells of 18-metre-long bowhead whales that help them repair their bodies. This could explain how they can live for more than 200 years!

Elephants are the largest land animals with the longest lifespan, reaching up to 70 years old. All that wisdom is passed down through the herd.

Tortoises are known for their old age. The oldest living land animal is a Seychelles giant tortoise named Jonathan, who was born in 1832, making him over 190 years old. Tortoises have a slow metabolism and their heartbeat is twice as slow as a human's. They can sleep for up to 16 hours a day.

The shortest lifespan of any known animal is the mayfly, which only lasts 24 hours. It's one of the oldest flying insects on Earth.

While larger animals have longer lifespans, there are some smaller animals (like bats, moles and birds) that live longer, too. These animals have special adaptations to help them, such as being sneaky or able to burrow into the ground away from predators. Smaller animals have a higher metabolic rate and generally have faster heart rates.

The ruby-throated hummingbird has one of the shortest lifespans of any bird. It only lives for 3–5 years, and has an incredible heart rate of 1,200 times per minute.

The average human lives to about 72 years old.

Fast growers like the pygmy shrew often tend to live shorter lives. While the shrew only lives for about a year, it can have about two litters of up to six babies in that time.

Pygmy shrews are the same size as a golf tee!

There is one animal which stays young forever: the immortal jellyfish. They have lived in the seas for over 600 million years, and they can even go back to their younger stage and repeat their life cycle.

Animals in colder places often live longer, since their heartbeat slows down along with the ageing process. The Greenland shark can live up to 500 years in the cold, deep waters, and the quahog clam – one of the oldest living things in the sea – is estimated to be 507 years old.

HOW DO ANIMALS EXPERIENCE TIME?

You know the phrase "time flies"? Well, for some animals it really does!

How different animals perceive time depends on how fast their nervous system reacts to events happening around them.

Fast vision and keen eyesight help animals detect rapid changes around them to help them catch prey or escape from a predator.

Smaller animals that need to avoid predators tend to see things unfolding more slowly, so that they can react quickly by hiding or running away.

Animals that move fast perceive time quicker than slow-moving animals. They have better senses to detect things happening around them.

In general, animals that are small or can fly perceive time the fastest, and flying animals detect light changes at a faster rate than animals that live on land.

Dragonflies can perceive changes in their environment the fastest, detecting 300 flashes of light per second. Flies see 250 flashes of light per second, making their world seven times slower than ours, which explains why it's so hard to catch one!

Small birds like the pied flycatcher have some of the fastest eyes around. Their vision is so fast that Usain Bolt's world-record-winning 100-metre-sprint would seem like a walk in the park to them.

Dogs are 25 per cent faster than humans in taking in visual information, which makes time move more slowly for them. One dog year is equal to seven human years.

Animals which are slow-moving or live at the bottom of the ocean tend to perceive time the slowest. Woodlice, for example, only see four flashes of light per second. The crown-of-thorns starfish, which lives on the seabed, perceives time amongst the slowest of animals. It sees just three flashes of light every four seconds.

HUMAN RECORD-BREAKERS

Humans are naturally competitive and we keep trying to beat the clock. Records are always being broken, but here are some of the most impressive ones!

- In October 2023, Kenyan Kelvin Kiptum ran the 26.2-mile Chicago Marathon in 2 hours and 35 seconds, which is a world record for male marathon runners.

- Ethiopian runner Tigst Assefa set the women's world record of 2 hours 11 minutes and 53 seconds in the 2023 Berlin Marathon.

- At the 2009 World Championships in Berlin, Jamaican sprinter Usain Bolt set the world record in the 100 m sprint at 9.58 seconds. That's faster than the average traffic speed!

- In 1988, American athlete Florence Griffith-Joyner set the women's world record for the 100 m at 10.49 seconds.

- In 2008, US swimming champion Michael Phelps swam the 200 m freestyle in 1:42:96 minutes, which is a speed of about 7.5 km/h.

- In March 2021, Budimir Sobat from Croatia broke the record for the longest time a person has held their breath underwater, with an incredible 24:37:36 minutes. That's longer than an episode of *The Simpsons*!

- The oldest confirmed person in history was Frenchwoman Jeanne Calment. She was born in February 1875 and lived to the age of 122 years and 164 days.

- In 2017, Slovakian Pavol Durdik set the record for most socks put on one foot in 30 seconds, with a whopping 28 socks!

- American Joey Chestnut has the record for being the fastest eater in the world, guzzling 141 hard-boiled eggs in 8 minutes, 102 tamales in 12 minutes, 182 chicken wings in 30 minutes, 76 hot dogs in 10 minutes, and much more.

Humans have also broken records when it comes to expeditions.

- In December 1911, Norwegian Roald Amundsen led the first expedition to reach the South Pole, beating Captain Robert F. Scott of the UK by five weeks.

- The Americans won the Space Race on 20th July 1969, when astronaut Neil Armstrong stepped onto the Moon for the first time with the famous words, "That's one small step for man, one giant leap for mankind".

- Russian cosmonaut Valeri Polyakov holds the record for the longest time spent in space, spending 437 days aboard the Mir space station from January 1994–March 1995.

APRIL
JUNE
AUGUST
DECEMBER
HAPPY BIRTHDAY!
24:37.53
45

THE END OF TIME

Will time ever stop?
What would the end of time look like?

We know time began with the Big Bang,
but some scientists believe that
the Universe could end with the Big Crunch.
In this hypothetical scenario,
the Universe stops expanding and starts
shrinking again. Everything could fall back
together to an extremely dense point,
like the reverse of the Big Bang.
Some scientists think that the Universe
is actually a cycle of Big Bangs,
expanding and shrinking over and over.

Another meteor could smash into the Earth and wipe out life like it did with the dinosaurs. What would happen then?

If humans went extinct, our power plants, sewers and electricity would shut down. Buildings would collapse and nature would take over our cities.

Just as our planet existed for over 4 billion years without life, it will last another 4–5 billion years without humans. Scientists estimate that the Sun will get bigger and brighter, heating up the Earth. In about 1 billion years, our planet will be too hot to support life, and will be a desert wasteland.

In about 5 billion years, our Sun will eventually run out of hydrogen. With nothing left to fuse in the core, it will expand into an even bigger star that astronomers call a red giant. This expansion will swallow up our planet and end time for Earth.

This is still billions of years away, though, so there's no need to lose any sleep over it! There's so much growing and evolving for us humans to do before then. What do you think will happen in the future?

INDEX